STRANGE
SUDDEN &
UNEXPECTED!

**True Stories from the files of the
Smithsonian Institution's Center for
Short-Lived Phenomena**

Written and illustrated by JAMES C.
CORNELL, JR.

SCHOLASTIC BOOK SERVICES
New York Toronto London Auckland Sydney Tokyo

Copyright © 1972 by James C. Cornell. All rights reserved. Published by Scholastic Book Services, a division of Scholastic Magazines, Inc.

1st printing ....April 1972
Printed in the U.S.A.

A WORD TO THE READER:

OUR SPACECRAFT EARTH

When the astronauts first walked on the Moon, they gave man more than he bargained for. Reaching for the stars, the astronauts found Earth.

In the void of space, Earth appeared as some huge blue-green marble set in black velvet. A beautiful sight, but also frightening. Suddenly, man realized that his planet was also a spacecraft on a long journey through time with only a limited supply of food, water, air, and natural resources.

This birds-eye overview of Earth created a new concern for our planet and for the wise use of its resources. The view from the Moon also revealed how fragile and small the world really is.

In this tiny, closed system that man calls home, every natural event, no matter how faraway or how insignificant, becomes directly connected to our lives. For example, a volcano erupting in Bali may contribute air pollution to the skies over Chicago. An oil spill in Florida may destroy fish that feed the people of Canada.

Even the discovery of a lost tribe of Stone Age men in South America is important to us, for their lives tell us how we evolved into a civilized society.

Gathering information about such far-flung and far-out events is the job of the Smithsonian Center for Short-lived Phenomena. This unique "news service for science" is helping man form his own overview of Earth without going back to the Moon.

In fact, right now, the earth looks like a pretty exciting place to be. Eruptions, earthquakes, animal migrations, meteorite falls, and all kinds of other incredible events happen every day.

No one knows exactly what will happen next. But, whatever it is, you can be sure it will be sudden, unexpected, and hard to believe.

James C. Cornell

Contents

FOUR MILLION AND
TWENTY BLACKBIRDS

FOUR AND TWENTY blackbirds baked into a pie
may be a tasty dish to set before a king; but *four
million and twenty* blackbirds in your backyard are
a nightmare.

Many Americans thought it a joke when nearly
four million cowbirds, grackles, and starlings de-
scended on tiny Scotland Neck, North Carolina, in
the early spring of 1969.

5

For the people of Scotland Neck, however, the blackbirds were no joke.

First, the birds made a horrible racket as they left their roosts in the morning and when they returned at night. Second, the droppings of four million birds have a vile and stomach-turning odor. Third, the lice and other parasites carried by the birds posed a serious health hazard.

When the birds didn't go away and when the droppings started to reach ankle depth and dead birds began to litter lawns, the people of Scotland Neck became frantic.

They even considered using some of the odd-ball suggestions for getting rid of the birds that came in from around the country. Like building giant scarecrows! Or shining bright searchlights on the birds so they couldn't sleep at night! Or playing loud rock music to drive the birds crazy!

One ingenious plan called for spraying the birds with detergent powder just before a rainstorm. Supposedly when the soap and rain washed away the protective body oils, the birds would freeze to death. The weather remained warm and clear in Scotland Neck, so the plan was never put in action. Perhaps it is just as well. No one ever considered what they would do with four million frozen blackbirds.

In desperation, scores of hunters went out and blasted away at the birds with shotguns. Only a relatively few birds were killed, however, and the rest simply flew back to the roosts when the hunters were finished firing.

Finally the birds just packed up and flew away. Apparently their stay in Scotland Neck was over and they were headed North to summer roosts.

But would they come back some other spring? Not likely, said the experts.

According to blackbird specialists, the half-billion American blackbirds usually migrate in great flocks of between 50,000 and one million birds. Although a flock may come back to the same general area each year, they rarely return to the exact same spot.

Well, anyhow, that's how the blackbirds are *supposed* to behave. Unfortunately the blackbirds who roosted in Scotland Neck didn't listen to the experts. Besides they seemed to like the place. And back they have come — year after year.

The return of the blackbirds has made Scotland Neck a minor tourist attraction. However, the townspeople would rather have peace and quiet.

Considering that every other anti-bird scheme has failed, the people of Scotland Neck now think they may have to cut down their trees so the birds will have no place to land.

OUT OF THE SEA

At FIRST, the pilot of the island-hopping airliner on route to Fiji from Tonga thought a ship was in trouble. He banked his plane sharply toward the faint plume of smoke rising off his starboard wing.

As he flew nearer the coral shallows known as Metis Shoals, he quickly realized this was no shipwreck.

A thick gray column of smoke billowed more than 3,000 feet in the air — straight out of the water.

Around the base of the smoke column, the sea

bubbled and boiled. And just beneath the surface, Metis Shoals glowed cherry-red.

A new volcano was being born before his eyes.

The Metis Shoals eruption soon became the most exciting event of the year for the islanders of the South Pacific. Journalists from as far away as New Zealand flew out to take pictures. Even the King of Tonga chartered a plane for a closer look.

The incredible sight was well worth the trip. The undersea volcano created a kidney-shaped island more than a half-mile long and 150 feet high. Curiosity seekers kept their distance, however, for huge molten boulders shot more than 1,000 feet out of the sea about once a minute.

By the time news of the volcano reached the United States, the eruptions had stopped and waves lapped over the new and now quiet island.

Unfortunately, the eruption produced a very fragile and frothy type of pumice stone that could not withstand the constant beating of the waves. Metis Shoals might soon disappear without a trace.

While undersea volcanoes are not unknown, they are extremely rare. Studies of their rock formation are vital to understanding how they are born, and scientists desperately wanted samples of Metis Shoals.

Luckily, Dr. Charles Lundquist of the Smithsonian Institution was on his way to Australia at the time. He detoured south to Tonga where he chartered a rusty old cargo ship that would take him and two native divers out to the vanishing island.

His little makeshift expedition arrived at the shoals just in time. Already, the fragile peaks of

Metis Shoals had washed away and only a few jagged points of rock poked above the surface.

While the ship circled at a safe distance, the two divers paddled out to the island and dove down to break a half-dozen basketball-sized pieces of lava from its submerged shores.

Shortly after Dr. Lundquist left Tonga, the island sank back into the sea. Today, at the Smithsonian in Washington, only those crumbly bits of gray rock remain to prove that Metis Shoals ever existed.

THE MISSING TRADER

JULIAN GIL was a brave — and sometimes reckless — man. A fur trader in the jungles of Columbia, South America, he placed his life in danger every time he ventured into the deep unmapped forests.

In December 1968, Gil and two of the local people set out on what would be his last expedition. Searching for jaguar, monkey, and deer hides, Julian Gil found instead a savage tribe of forgotten Indians.

According to later reports, the Gil party had a

11

long and difficult trip into the jungle. After days of trudging through neck-deep swamps, over rugged trails, and across snake-infested rivers, they finally stumbled on a village of some 200 natives all living in a single *maloca,* or huge cone-shaped hut about 80 feet in diameter.

The people spoke an unknown language and used tools straight out of the Stone Age. Gil must have realized that these were people that few, if any, white men had seen before.

Perhaps hoping to establish the first trading post in this untouched wilderness, he made camp among the Indians. One guide remained with him, the other headed back to the town of La Pedrera with instructions to send out a search party if Gil had not returned within two months.

Julian Gil was never seen again.

At the end of the two months, Julian's brother Efrain organized a search party. Accompanied by a detachment of marines from the Colombian navy, Efrain traced his brother's path back to the tiny village.

Although the village seemed deserted, the men felt sharp eyes watching their every move from the bushes. While the marines fanned out to search for the tribe, Efrain Gil poked through the rubbish of the camp for signs of his brother. He found a belt and buckle that belonged to the native guide and a necklace made of buttons from Julian's shirt.

In the meantime, the marines captured a group of natives and brought them into the hut as captives. Throughout the long night, the captured Indians wailed and cried, calling out for help to their tribesmen hiding in the forests.

Spared from attack that night, the entire party left at dawn for the long trek back to civilization. They were followed closely by angry natives.

The Indians grew bolder with every mile. Suddenly, the Indians made a move as if to attack, and the marines opened fire with automatic rifles. In the confusion and killing that followed, some of the Indians escaped; but one man, one woman, and four children — apparently all members of the same family — were kept as hostages.

When the party finally reached the coast, news of their contact with an unknown tribe spread swiftly to anthropologists around the world. Experts from several universities attempted to communicate with the Indians, but with little success. Even Indian interpreters speaking 15 different dialects couldn't understand the primitive language.

According to one anthropologist, the natives might be survivors of the Yuri tribe thought to have disappeared more than half a century before. The Yuris once numbered in the thousands, before Colombian slave traders killed off most of them at forced labor in the Colombian rubber plantations early in this century.

The only true way to link the Indians with the lost Yuris would be to live with them and compare their current language with the crude word-lists drawn up by slavers in the 1800's. Unfortunately, any further contact with this tribe — and any determination of its primitive life style — is probably impossible now.

The episode with the Gil brothers and the marines probably has left the Indians frightened, hostile, and wary of white men. Any exploration of this area will be extremely dangerous for years to come.

Indeed, in July 1969, when the family of hostages was returned to their jungle village, the huge hut stood deserted and silent, stripped of any sign of life. Released by their captors, the six Indians disappeared into the jungle. No one has seen them — or any members of the lost tribe — since.

STARS FELL ON ALLENDE

THE CHURCH BELLS in Chihuahua, Mexico, had just struck one in the morning as Guillermo Asunsolo, editor of the *El Heraldo*, put his last edition to bed.

Suddenly, a bright, almost blinding, light turned night into day over Chihuahua. Reporters and pressmen rushed to the windows thinking they might see their last "big story" — the end of the world.

Instead, they saw a gigantic fireball, brighter than the full moon, streak across the sky. The fireball — really an extremely bright meteor or falling star — appeared to strike earth somewhere in the rugged Sierra Madre mountains.

As the fireball disappeared to the South, the newspaper building shuddered under the force of a massive shock wave.

Almost immediately, phones began to ring with calls from the mountain villages of North Central Mexico, where the fireball had been seen as a sign from God. The wire service teletypes started flashing too, with reports of observations as far north as El Paso, Texas, and Tucson, Arizona.

Like any good newspaperman, Asunsolo called for a special edition to report this wonderous sight. Little did he know that the biggest story was still to be written.

The next morning, villagers in Pueblito de Allende, a little town south of Chihuahua straddling the Pan-American Highway, woke to find their yards, fields, and streets speckled with thousands of black-crusted meteorites.

The fireball had dropped a veritable shower of stones ranging from pebble- to football-size on their town. One large meteorite had fallen not fifteen feet from the post office.

American and Mexican scientists, including a team from the Smithsonian, immediately rushed to Pueblito de Allende in hopes of recovering meteorites, cataloging their types, and determining the extent of the shower.

Freshly fallen meteorites are extremely valuable to astronomers. As leftovers from the formation of

16

the solar system, these pieces of cosmic debris hold many clues to the mystery of the universe.

The scientists used a very effective technique for gathering samples. They set up a soft drink stand on the edge of the town and offered Indian children a free drink for every meteorite they returned.

During the next two weeks, the scientific teams recovered more than two tons of meteorites, thus making this the largest fall on record. The meteorite hunters feel another two tons may have gone unrecovered — at least by science.

For the Indians of Mexico's high country, these stones from the sky had magical properties. Many of the missing meteorites are now carried as good luck charms and as reminders of the night when the stars fell on Allende.

OF MOOSE AND MEN

THE FOLKS in Alaska like to see wild creatures running through the woods and over the tundra. After all, that's why they moved to America's "last frontier" — to be close to nature!

But even the most devout nature lover can get too much of a good thing. Especially if it means a herd of moose in your backyard!

In the winter of 1969, homeowners in Anchorage, Alaska, complained to game wardens about the moose on their lawns. All around the city, large herds of moose gathered to nibble at tree trunks, forage for garbage, or just stare at startled Alaskans.

Although moose are found throughout America's wildest state, the sight of so many so close to the city was rare. Usually the large animals prefer to keep their distance from man, partly out of fear and partly out of dislike.

The Alaska Fish and Game Department told the Smithsonian that the sudden appearance of the moose in town actually had been building up for years.

The moose population had multiplied rapidly during several recent years of relatively mild weather and plentiful food. Moreover, a change in the hunting regulations had provided more protection for moose cows, so many new young moose had been born.

At the same time as the moose population was growing in the wilderness, the human population in the cities of Alaska was expanding. Anchorage was booming and the first signs of a "suburban sprawl" had appeared. As new homeowners built their split-level dreamhouses farther and farther from the center of the city, they started to overlap on the old wintering grounds of the moose.

Then, in the severe winter of 1969, bitter cold and heavy snows drove the moose down from the slopes of the mountains into the warm coastal regions near Anchorage. And moose and man finally came face to face.

Actually, most Alaskans took it all in stride. In fact, the big-game hunters immediately petitioned the Fish and Game Department to extend the moose season and increase the bag limits.

For once, hunters and conservationists reached agreement. Selective hunting (or "harvesting") of the moose herds seemed to be the most humane action, for it would reduce the herds to levels that could survive on the available food and it would also reduce the strain on the homeowners.

Of course, there was another solution, but no one suggested it. The Alaskan suburbanites could stop building homes on the moose's grazing grounds.

YUNGAY NO LONGER EXISTS

AT 3:34 Sunday afternoon, May 31, 1970, most of the 20,000 people in Yungay, Peru, were huddled around their radios listening to the last exciting moments of a World Cup soccer game broadcast from Mexico City. Few even bothered to stir when the low grumble of an earth tremor shook their homes.

Earth tremors were common in Yungay. The high

21

narrow valleys of the Peruvian Andes often suffered as many as 1500 a year!

But the rumbles this day were deeper and stronger than ever before. Suddenly, the whole world seemed to tip upside down and the voice of the radio announcer was drowned out by the sound of crashing walls and human screams.

Then, just as suddenly, it was over and Yungay lay in ruins. Houses had crumbled, the streets had buckled, and the central plaza had split open. The survivors clasped each other and gave thanks for their salvation.

Perhaps they gave thanks too soon!

Ten minutes later, another low rumble was heard — this time from high above their heads. Huge chunks of ice and stone shaken loose by the quake rolled down the barren mountain sides in a deadly avalanche.

One final and almost unbelievable horror was yet to come. As the boulders and ice tumbled into high mountain lakes and reservoirs, dams and retaining walls burst spilling millions of gallons of water down the valley toward Yungay.

A solid wall of mud, many feet deep, thundered down the funnel-like valley, covering everything in its path. Within mere seconds, the beautiful city of Yungay and most of its 20,000 inhabitants disappeared from the face of the earth.

Terrible destruction occurred throughout all of northern Peru in the wake of this, the worst earthquake in the history of the Western Hemisphere. Almost 70 percent of Chimbote, a major seaport, was destroyed. Scores of other large cities had similar damage. And untold numbers of tiny Indian villages — most unnamed and unmapped by the government — simply vanished forever.

Still, the worst horror was in Yungay's valley. From the air, it looked as if some giant had filled this natural crack in the earth with brown plaster. The only traces of Yungay, a city founded four hundred years earlier by the Conquistadors, were the tops of four royal palm trees that once marked the city square.

No one knows exactly how many people are buried under the mud. Perhaps someday future archaeologists will uncover the remains of these lost people, just as today's diggers turn up the bones and relics of the ancient Incas, and as yesterday's diggers uncovered Pompeii and Herculaneum.

THE PLANT OF
MYSTERIOUS DEATH

IN HUNDREDS of old horror movies about haunted mansions in the Deep South, ghostly images of Spanish moss draped like death shrouds from decaying balconies and tree branches have provided a backdrop to terror.

Now, the Spanish moss — so often associated with mysterious death — is mysteriously dying itself.

Really neither "Spanish" nor a "moss," the plant

is a very distant cousin of the pineapple with long stringy gray-green leaves and tiny yellowish blossoms. Found throughout the South from the Gulf Coast to Virginia, as well as in Central America and on the islands of the Caribbean, Spanish moss has often been falsely accused of being a killer itself. In fact, some people call it the "Vampire of the Forests."

The poor plant gained this rotten reputation because it is usually found festooning dead trees and people wrongly assumed that it sucked the life from other living plants. Actually, the plant is no killer. It takes nourishment from moisture and particles carried on the wind and simply clings to any structure — including light fixtures and telephone poles — that will provide a convenient anchor.

Ironically, *something in the air* now threatens to destroy this familiar symbol of the Old South. An unknown disease is killing off the moss everywhere.

Should the Spanish moss ever disappear, the South would lose the source of some very colorful legends. For example, the Indians claim the plant grew from the hair of a maiden who killed herself for the love of a young brave. Her long tresses have been draped over trees ever since, say the Indians, as a tribute to her lost lover.

The French explorers in Louisiana called the plant "Spanish beard" as an insult to their competing explorers in the New World. The Spanish returned the insult by dubbing the plant "French hair."

The early settlers — who finally stuck with the name Spanish moss — mixed the stringy plant with mud to make mortar for their cabins. They also used cured and dried moss to make bridles, horse collars, and saddle blankets. In the early part of this century, enterprising Southerners harvested the plant with knives tied to the end of long poles. They sold the fast-growing plant as stuffing for mattresses and padded furniture before the invention of synthetic padding materials.

Someday soon, perhaps, the Spanish moss could become just a legend itself. Unless, of course, science can find the cause — and the cure — for its mysterious malady.

The plant is usually resistant to all sorts of bugs, so scientists think the cause might be some new type of exotic plant virus. Other experts, however, think the answer may be more simple.

Because Spanish moss literally "lives on the air," one likely culprit for its killer is modern air pollution. If man doesn't clean up his air soon, the mysterious and ghostly Spanish moss may truly become a ghost.

DEADLY SUPERBUGS

CAN YOU IMAGINE being driven from your home by bugs?

While ordinary insects may never force you to give up everything and run for your life, just hope you never meet a South American "superbug" called the *chirimacha*.

The chirimacha is an insect pest four times larger than the common housefly. It thrives on garbage, is immune to most pesticides, and can live for months without either food or air.

Worse yet, the chirimachas carry Chagas' Disease, a rare tropical illness that is almost always painfully fatal. Victims of Chagas' Disease suffer intense headaches, nausea, fainting, and nosebleeds before finally wasting away in a slow and terrible death. At present, there is no vaccine against — or cure for — this killer disease.

No wonder then that the people of the Vitor and Camana valleys of Peru were frightened out of their minds. For several years, this region of Peru was infested by literally millions of the ugly chirimachas.

Although the insect is often found as a parasite on small animals, it really prefers to live in human kitchens where it can feed on table scraps and garbage. Indeed, the insects reproduce at a fantastic rate amidst the dirt and filth of crowded slums.

The Peruvian government made several attempts to disinfect the crowded neighborhoods invaded by the bugs, but it did little good. The slum dwellers were no more successful with their homemade efforts of pouring boiling water and burning kerosene on the bugs.

Finally, many families simply abandoned their homes and fled to the cooler safer ground in the mountains. The chirimachas prefer the warm humid valleys and there they stayed, controlling the ramshackle huts and tenements and threatening the population with a terrible epidemic of Chagas' Disease.

Of course, if housing and sanitation were improved and the slums eliminated, the bugs probably would disappear. Unfortunately, the Peruvian government has taken few steps to improve the living conditions of the poor. Instead, the slum dwellers must continue their losing battle against the deadly "superbugs."

THE SLOW
EARTHQUAKE

For CENTURIES the people of Naples, Italy, and surrounding towns have lived in the shadow of death. The sleeping Mt. Vesuvius across the bay is a constant reminder of the potential turmoil just beneath the earth's crust.

As if that wasn't enough, a part of the Pozzuli slum section is known as the Campi Flegrei, or "fiery fields," and smoke and steam continually seep from ugly cracks in the soil.

For more than a decade, Pozzuli also has been slowly rising out of the ground. In the spring of 1970, the rising suddenly accelerated and the streets and buildings lifted as much as three feet in some places.

Known to scientists as the "Pozzuli Uplift," this strange phenomena might be better described as a "slow earthquake."

Slow or not, the uplift caused serious destruction. Old rickety buildings suffered the most, as the shifting of foundations crumbled mortar and cracked stonework.

The worst problem, however, was the wave of panic and terror that swept through the grimy neighborhoods of the slum.

Daily reports from Pozzuli claimed that new volcanic cracks had appeared in the fiery fields and that lava was oozing out. Jittery citizens reported earth tremors almost hourly.

Fishermen returned to Naples harbor with wild stories of finding cooked fish in blackened nets burned by undersea eruptions.

None of the stories — earthquakes, cracks, or cooked fish — could be verified by government officials. But this didn't stop people from panicking. Thousands fled their homes and angry editorials demanded government aid.

Pozzuli had disaster on its mind and nothing could reduce the fear of being trapped in the packed tenements and narrow streets if the buildings collapsed.

The fear of Pozzuli's people was not entirely foolish or irrational. Living in Pozzuli is much like living on the lid of a boiling pot. As the heat under the pot increases and decreases, the water boils, bubbles, and cools so that the lid rises and falls with the heat.

Some day — perhaps next century, or next year, or even tomorrow — the internal heat of the earth's burner may suddenly turn up to the point where the pot boils over and the Pozzuli lid blows sky high.

BEACHED BLUBBER

THE FEW hardy souls who braved the unusual cold spell to stroll along the beach at Ft. Pierce, Florida, one morning in January 1970, saw an unusual sight.

More than 150 whales — an entire herd — were lying high and dry on the sand.

The black whales, most of them more than 20 feet

long and weighing more than a half-ton each, were strung out along a two-mile stretch of beach front.

Many of the whales were still alive and making those strange, sad, squeaking noises scientists think they use for communication.

Similar strandings of whales have been reported from other parts of the world, but this was the first time so many had ever been seen on a Florida beach.

Men from the Florida Department of Natural Resources managed to loop ropes around the tails of some 35 bull whales and pull them back into deeper water. But, almost as soon as the whales were released, they spun around and headed back toward the beach faster than the men in boats could head them off.

A second towing attempt with another 50 whales was more successful. And one 12-foot female was taken from the beach and trucked to Marineland at St. Augustine, where she was nursed back to health. The other whales were not so lucky. They died on the beach from dehydration.

Why had the males come ashore at Fort Pierce? Even marine biologists were stumped by the apparently suicidal behavior of the huge aquatic mammals. Sometimes whales will beach themselves when they are sick and about to die. But blood samples from the Fort Pierce whales showed they had nothing physically wrong.

The scientists finally decided the beaching of the whales must have been the result of a social error.

Whales travel in herds led by an old bull — much like cattle does. Perhaps the bull himself had been sick and headed for the beach. Or perhaps he simply wanted warmer waters (the temperature was about

26 degrees that day) and accidentally strayed into the shallow water along the shore. No matter what the reason, the old bull must have headed into shore and his herd followed right behind.

Once in the shallow water, the herd probably ran into serious problems. Whales navigate and communicate by sending out sonorlike signals similar to the underwater radar used by submarines. Unfortunately, this signal becomes weak and muddled when aimed toward the type of sloping, sandy beach found at Ft. Pierce.

Unable to navigate in the shallow water, the whales may have become confused and frightened. Herd instinct probably took over, and the whales rushed headlong onto the beach and died.

The hysterical behavior of the whales is very much like that of the stampeding cattle who run over the edge of a cliff in their fright and panic.

TSUNAMI MEANS
DESTRUCTION

NOTHING is so feared by the people living on the islands or coastal regions of the Western Pacific as the dreaded *tsunami*.

The tsunami (pronounced sue-na-mee) is a Japanese word that means "big wave." For the people of Asian lowlands, tsunami also means death and destruction without warning.

Understandably, then, one of the few reports of a natural disaster to reach the Western world from

Red China in the past twenty-five years concerned a giant tsunami that nearly wiped out Shantung.

On April 23, 1969, a 20-foot tidal wave, whipped by 200-mile-per-hour winds, swept ashore in the great plains area where the Yellow River empties into the Gulf of Pohai between China and Manchuria.

Knocking over retaining walls as if they were paper cups, the waters surged 13 miles inland all along a 45-mile stretch of coast.

According to one report out of Hong Kong: "The winds and waves smashed houses like matchsticks, damaged larger buildings, and destroyed 1,000-square miles of rich farmland." At some points as far inland as 15 miles, the water was three feet deep.

Asian oceanographers claimed the big wave was the worst to hit China in more than 80 years. The Shantung peninsula is one of the most densely populated areas in the world, so that more than 100,000 people were made homeless by the massive flooding.

The loss of homes and factories and food was all the more disastrous because subzero weather and snowstorms hampered the efforts of soldiers and civilians trying to reach the stranded people.

Ironically, the disaster may have helped cool off China's internal political problems. In the months before the tsunami, China had been boiling with a civil strife between the pro- and anti-Mao forces. All the political differences were forgotten as national resources were mustered in a single-minded relief effort.

Chinese officials claim that millions of militiamen

and soldiers poured into Shantung to aid the stricken people. After the massive clean-up and rescue operation, the internal struggle seemed to slacken and Mao came back in full control.

Even the Chinese Communists, who claim to have conquered almost every other force in the world, must bow before the terrible power of the tsunami.

INVASION OF THE BALLOON SPIDERS

"Unusual flying/floating material observed near St. Louis at McDonnell-Douglas Space Center. Bubble-shaped, filmy material drifting northward. Objects ranging from dime-size to 10-foot globules sighted."

When you operate a Center for Short-Lived Phenomena, you often receive all sorts of far-out reports about all sorts of far-out things. But this one was really far-out! In fact, it sounded like the typical "creatures-from-outer-space story."

"Samples retrieved appear to be fibrous material, pure white and sticky when in contact with grass, metal, or other objects. No cell structure visible even under an 800-power microscope."

What in the world could this be? Some horrible new kind of air pollution? The messy fall-out from some industrial smokestack?

Or, was it something from out of this world? The debris of an interstellar spacecraft that had exploded over our planet? Sticky gelatin bombs dropped by extraterrestrial invaders?

Actually, the strange white bubbles turned out to be made neither by men on this planet nor supermen on another. The weird floating fibers were the products of one of nature's oddest forms of reproduction.

The city of St. Louis had been invaded not by Martians, but by balloon spiders.

The answer to the puzzle came from biologists at the county health department. They found the material looked much like the thin strands of fiber spun by silk worms. More important, in one globule, they found a small brownish-yellow spider about the size of a penny.

The balloon spider is peculiar among the insects for its strange mating habits. The female climbs to the top of a tall tree where she spins a filmy circular web. Into this web, she deposits her eggs. Then she cuts the web loose so the wind can carry it off. Wherever the breeze carries the balloonlike web and its load of eggs, a new colony of spiders will spring up.

Farmers in Missouri are quite familiar with the

drifting webs. But city dwellers seldom see the phenomena. On the day of the St. Louis invasion, however, the wind and weather conditions were perfect for carrying the webs out of the country and into the downtown streets.

At the height of the "invasion," webs filled the air with a density of almost one per square foot. For city folk unaccustomed to nature's oddball ways, it was unbelieveable. No wonder they thought the spider invasion was something from out of this world.

THE GREAT GASSER

IN THE FALL of 1969, the Yugoslavian company Naftgas was boring a deep well near the little town of Becej on the Hungarian border. These modern prospectors were seeking natural gas, the prime source of heating and cooking fuel for many parts of Europe.

To their dismay, the Naftgas prospectors found much more gas than they could handle.

Suddenly and without warning, the drillers broke through an underground crystal block. The drill spun wildly for a few seconds, then a tremendous explosion lifted the drilling rig off the ground and a geyser of air shot out of the hole.

The drillers had inadvertently tapped a small gas volcano.

The manmade blow-hole emitted high pressure air for almost six months. The worst part of the geyser was the horrible whistling noise caused by the escaping air. It sounded like a million screaming banshees.

The villagers of Becej had finally become accustomed to the terrible sound, when it suddenly stopped. For one day — April 9, 1970 — an unnatural silence fell over Becej.

Officials from Naftgas gingerly inspected their creation. Could they start drilling again? Had the volcano stopped? Hardly!

Again without warning, the well suddenly exploded. This time the eruption sent great clods of earth hurtling hundreds of yards into the air. A large dirty cloud rose over the well. When the smoke and dust cleared, the drillers and villagers found a huge crater more than 100 feet wide. And out of this crater seeped the noxious deadly natural gas — 93 percent carbon dioxide and seven percent methane.

That night, five people in Becej died from gas poisoning, another 10 were hospitalized. The government ordered the town evacuated and the well capped, if possible.

Workers poured thousands of gallons of water into the well to break down the gas. Huge motor-driven fans were set up to purify the air. Still, the crater continued to rumble and grumble for almost two months. Occasionally large hunks of earth shot into the air like mortar shells.

Then, on June 4, the eruptions stopped once more. This time, however, the calm lasted only 30 minutes. An ear-splitting explosion rocked the town and a piece of earth as large as a small building shot from the crater's mouth.

But this was the geyser's last gasp. The walls of the 400-foot-deep crater fell in on themselves, sealing off the gas flow forever.

Today, there are few drilling operations around Becej. No one wants to create another manmade volcano.

A BLACK CHRISTMAS

On Christmas morning 1969, children living around Lake Vattern in southern Sweden looked out the windows and began to cry. As far as the eye could see, the white mantle of Christmas snow had turned jet black.

Fresh snow had fallen on Lake Vattern for several days before Christmas. Then, on Christmas Eve, the snow had turned to a light, freezing rain.

Now, on this Christmas morning, the bright, gay,

winter landscape had turned dark and dull. Every inch of the ground was black and grimy, just as if evil elves had painted it, playing a cruel joke.

Deer and rabbits running through the fields seemed as puzzled as the children. When the animals ran over the snow, their hoofs and paws turned up white tracks. Apparently, the black snow was only on the surface and had probably fallen with the rain.

Whatever it was, the black material was certainly gritty, and almost impossible to remove from clothing, even with strong detergents. Tracked inside, the black snow melted into little shiny pools of oily liquid.

One of the most famous writers about the mysterious ways of nature, Charles Fort, once compiled an entire book devoted to unexplained and mysterious "rains and snows." Fort claimed to have seen falls of red snow, fish, flowers, and even frogs. Unfortunately, few of Fort's amazing incidents were ever studied or documented by reliable eyewitnesses or scientists.

Today, Fort's stories remain only oddities, more fiction than fact. But there is no fiction about Sweden's black snow of 1969.

A team of scientists from Stockholm's Ecological Center rushed to Lake Vattern immediately for a first-hand investigation. They analyzed the black snow — and found it contained dangerously high levels of DDT as well as a host of other industrial pollutants.

Unlike that mysterious "rain of frogs" reported by Charles Fort, there was no question about this snow fall. It was the work of man.

Sweden's black snow was still another example of how industrial pollution is ruining the atmosphere and destroying the beauty of our planet. Indeed, for the children of Lake Vattern, pollution even stole Christmas.

SQUIRRELS ON
THE MARCH

MOTORISTS traveling the highways of the East Coast in September 1968 were the first to notice something strange was happening to the gray squirrels.

Almost 1,000 times as many dead squirrels as usual littered the roads. The sight of dead animals on our high-speed expressways is common, but when unusual numbers of bodies appear on the asphalt, biologists know something odd is happening in the bushes too.

Farmers, hunters, and game wardens confirmed these suspicions. Throughout the Appalachian region of Maryland, Virginia, Tennessee, and the Carolinas, thousands of gray squirrels were seen scurrying through the underbrush, moving across open areas, invading suburban gardens, and even swimming in reservoirs and rivers.

Typically, the nation's newspapers overreacted to the story and distorted the facts. According to press reports, "gigantic hordes of hungry squirrels were headed north in a mass migration caused by disease, starvation, and insanity."

Many well-meaning and soft-hearted Americans were so moved by the plight of the "hungry and tired" squirrels that they launched campaigns to feed them. Supermarket posters in North Carolina urged people to establish feeding stations. People sent checks to the state Fish and Game Departments to pay for squirrel food. And one group of Tennessee citizens even appealed to the people of Florida to gather acorns for shipment North.

Less kind-hearted, but more practical, hunters urged state game wardens to open the squirrel season early and double the bag limits.

Professional biologists found the public reaction funny, but they were no less fascinated by the strange squirrel behavior. The behavior was even more mysterious when all the popular theories about its cause proved false.

For example, almost all the squirrels had bellies full of black cherries and acorn mash, so they certainly weren't hungry. Nor did they seem bothered by sickness or even parasites that could have driven them nutty.

In fact, the scientists couldn't even find any indication that the squirrels were traveling in flocks or headed northward. Most of the squirrels seemed determined individualists, traveling alone — and headed in every direction imaginable.

The biologists did find, however, that the squirrel population was unusually large that fall. The strange behavior was related to this overpopulation — but not because they had been driven crazy by overcrowding.

Similar squirrel "migrations" had been recorded before. In the days of the last century, before the great forests had been cut down, they were quite common.

Strangely enough, all the migrations also took place in September. Why did the squirrels start moving about so wildly at harvest time when the food was the most plentiful?

The answer to this puzzle is found in the peculiar habits of the gray squirrel, the type of squirrel found in most big city parks.

Unlike his cousin the red squirrel, the gray squirrel doesn't store his food in a hollow tree. He simply tucks nuts and seeds under leaves or a few inches of dirt. This storage technique naturally uses up a lot more territory than if he simply picked one central spot.

Each fall, the gray squirrels go through a major "reshuffle." As they run about looking for acorns and new and different spots to stash them away, the squirrels move farther and farther away from their regular haunts. Sometimes they even invade back-yards or swim across rivers in search of new hunting and hiding grounds.

Away from their regular haunts, the squirrels easily become confused and disoriented when frightened. In short, they act crazy.

The so-called "migrating squirrels" were really doing what comes naturally. However, there were so many more of them in the fall of 1968, they caught the attention of people usually not interested in squirrel behavior.

Biologists warn that more of these "migrations" may be on the way. The growing concern for ecology and conservation is helping to preserve and create more forest land, and, in turn, more squirrels. Those furry travelers with the bushy tails may be a common sight in the Septembers of the future.

CORNELL

PREHISTORIC STEW

BESIDES TOUCHING OFF student strikes across this country, the invasion of Cambodia by American troops in the spring of 1970 also may have helped speed the extinction of an unusual animal.

Amidst the turmoil and confusion of the Cambodian affair, biologists in Southeast Asia announced that the last of the "koupreys," a rare species of wild cow, had probably disappeared. The kouprey had become still another victim of the long Indochina War.

Before the expansion of the war, only some 50 to 60 koupreys still existed in three different areas of Cambodia. No koupreys lived in captivity, for all attempts to capture and breed failed.

Many biologists described the kouprey as a "living fossil," perhaps the missing link between prehistoric cattle and the present-day hump-backed Brahma bulls of India.

Studies of the kouprey bones at Harvard revealed the cattle had teeth almost identical to those of the "proleptarbos," a long-extinct prehistoric ox. Many scientists believed the kouprey had remained virtually unchanged for some 100,000 years.

Naturalists wanting to study the kouprey for other prehistoric links hoped to raise one or two herds in captivity. But these hopes were dashed by the intensified fighting in Cambodia. Two of the areas where kouprey grazed lay close to the Viet Nam border and had long served as hideouts for the Viet Cong. However, the third area had not been infiltrated by guerrillas until the sudden escalation of the war.

The allied drive into Cambodia forced the Viet Cong to retreat into this last kouprey sanctuary. Because of the heavy fighting and scarcity of supplies, government officials feared the Viet Cong might kill the kouprey for food.

After the fighting died down, searches of the area found no sign of the rare cattle. The search continues, however, in the hope that a small herd might have escaped into the forests to breed again.

Sadly, it is more likely that the kouprey's 100,000 years of natural history became hot stew.

CORNELL

CATCH A FALLING STAR

GUNTHER (SKIP) SCHWARTZ is one of those rare humans who know how it feels to catch a falling star.

On Saturday night, January 3, 1970, Skip sat watching television in his home in Lincoln, Nebraska. Suddenly, the news announcer interrupted with the report that an incredibly bright fireball had just streaked over the Midwest, spewing sparks and smoke behind it.

Most other men would simply have switched off the television and gone to bed. But Skip Schwartz bolted from his chair and began calling television stations and airports for more information.

The reason for Skip's unusual reaction to this news is his unusual job. Skip is the field manager of the Smithsonian's Prairie Network, a system of 16 automatic camera stations spread over seven states to photograph fireballs (super-economy-size falling stars) and to help recover meteorites (the extraterrestrial material that creates the fireball's light and sometimes falls on earth as space debris).

The eyewitness report of an astronomer in Kansas indicated that any meteorite from this fireball might have fallen somewhere in either southeastern Kansas or northeastern Oklahoma.

Skip immediately called the camera station caretakers in these areas and asked that films be sent to network headquarters in Lincoln. When the films arrived and were developed two days later, they showed a brilliant fireball trail, brighter than the full moon and lasting some 9 seconds. Using a computer to compare photos from two different stations, scientists plotted the fireball's path to the ground and predicted the possible impact point to be near the tiny hamlet of Lost City, Oklahoma.

Skip bundled himself and a pile of maps into a truck and headed South to Oklahoma. He hoped this wouldn't be another wild goose chase. Most fireballs burn up completely. Even if a meteorite does fall on the ground, it is like looking for the proverbial needle in a haystack. After five years of photographing the sky and searching the ground, the Smithsonian had not yet recovered a single meteorite.

But this time would be different!

At the start, the situation looked as bad as usual. A nine-inch snowfall covered the ground just before Skip Schwartz arrived in Lost City. It was two days before the weather cleared and he could venture out on the lonely roads.

On one such snow-covered road, only a half-mile from the predicted impact point, Skip drove over a black rock about the size and shape of a Vienna breadloaf. Something about this rock made him stop and pick it up. Just one look and he became hysterical with joy.

He had accidentally driven right into the object of his search. This heavy, dark-crusted, smooth-surfaced stone was the Lost City Meteorite!

Of all the hundreds of meteorites found by scores of men in tens of centuries, the Lost City Meteorite is one of the most valuable. Not only did the quick recovery time (less than a week from fall to lab!) make possible delicate measurements of radioactivities created in space, but the photographic record of its fall even told scientists what part of space it came from.

More important for Skip Schwartz — he had finally caught his falling star.

THE GREAT FROG WAR

FOR SIX DAYS the battle raged through the jungle in Southeast Asia. More than 10,000 warriors tore into each other, killing and wounding hundreds in a battle that swung back and forth.

A clash between the allies and Communists in Viet Nam? A guerrilla raid on some government outpost? A tribal battle?

Not really. The fierce warriors in this strange battle were 10 different species of frogs and toads apparently fighting to the death.

"The Great Frog War" began on November 8, 1970 as a small skirmish between about 50 frogs in a jungle clearing near the Malaysian seaport of Sungai Siup. The numbers quickly swelled, and, by the end of the day, some 3,000 frogs were biting and ripping at each other.

The next day the battle resumed in the shadow of an ancient Hindu temple. The swamp soon became a swirling mass of green, yellow, brown, and gray bodies, all clawing and tearing and making horrible sounds. Residents of the area who gathered to watch the strange fight claimed the frogs even carried off their dead and wounded.

Other smaller fights continued for almost a week. Eventually, an estimated 10,000 frogs and toads joined the battle. Local natives claimed the war was an annual event, with the frogs fighting over choice feeding and breeding grounds.

However, scientists from the Zoology Department at the University of Malaya had another explanation. In a report to the Smithsonian Center, the Malaysian biologists told of visiting the battle scene and finding the swamp waters full of frog eggs and tadpoles.

This had been no "war," they said, but rather an amphibian sex orgy!

The frogs simply were doing what came naturally — mating during a rainy spell that had followed a long drought. Unfortunately, the noise and activity of the mating attracted a species of toad that secretes a poisonous substance from its skin fatal to frogs.

Sometimes, too, as the big bullfrogs fought for the attentions of females, they jumped at each other, biting and tearing with their claws. Some frogs also had abrasive skin on their chests that ripped the hides of the others.

Despite all the appearances of a deadly battle, the frogs really were *making love, not war.*

The so-called "Frog War" had an odd sidelight. The superstitious people of Malaysia think the frog wars are bad omens of terrible disasters to come. In 1969, just 12 days before the bloody Malay-Chinese race riots that killed thousands of people, frogs had battled in a swamp near Butterworth, Malaysia. In years past, other wars supposedly had preceded floods, storms, and major revolutions.

"Pure nonsense," said most scientists, discounting any relationship between the frog wars and other calamities.

Yet, just one day after the "Great Frog War of 1970" ended, a gigantic cyclone roared up the Bay of Bengal and struck nearby East Pakistan killing a half-million people.

Only a coincidence? Who knows?

A BROWN WAVE OF ANTS

Jose Castellanos worked a tiny farm in the
Lares Valley of Peru. This lush, green land is well-
known for its tea- and cocoa-leaf production. It is a
rich and beautiful valley.

Jose was proud of his small plot of land and its
bountiful crop of cocoa plants. At harvest time, he
happily cut his own crop and carried it to market.
His farm provided well for his small family.

One morning in early 1969, Jose awoke to hear a strange, rustling noise, almost as if a brush fire was burning downwind in the valley. Jose jumped from his straw pallet and rushed outside. He saw no smoke, but the weird noise seemed to be coming closer.

He climbed a small knoll behind his hut and peered into the early morning mist toward the sound. Jose choked with fear, for what he saw was worse than any fire.

Slowly moving down the valley toward his cocoa plants was a great, coffee-colored wave. Millions upon millions of giant brown ants were sweeping down the valley like some mammoth lawn mower.

Behind the ants, the green vegetation of the valley — the precious tea and cocoa plants, the fruits and vegetables, even the green grass of lawns — had disappeared. In their wake, the ants left a beige and barren wasteland.

The ants that destroyed the Lares and La Convencion valleys of southern Peru in 1969 were known as the *Coqui*, a species of destructive leaf-cutting ants.

This type of ant has been a traditional scourge of farmers in the tropics, but rarely did they attack in such numbers as this year. Every farmer in Peru's valleys prepares for these insects, but none were ready to face the millions this time. The sheer number of ants in this "invasion" made control virtually impossible.

Worse yet, these ants had a special fondness for the tea and cocoa leaves that were the pride of the Lares valley. An entire tea or cocoa plant could be denuded within seconds by the hungry insects.

Eventually, after destroying thousands of dollars worth of crops, the ants turned away from the cultivated land and headed for the open fields. Eventually, too, simple overpopulation killed the ants, for the countryside simply could not support their enormous numbers.

Although many farmers were ruined by the invasion of the ants, the situation could have been much worse.

In some parts of South America, there are species of ants that eat animal flesh as well as plants. Hordes of army ants have reportedly eaten entire forests, including any animals — and humans — unfortunate enough to be caught in their path.

THE SLIDING GRAPES

OLD MRS. UDVARI lived high on the steep bank of
the Danube River near the Hungarian city of Dunafoldvar. With a few chickens and a pig, she lived a
simple life of few luxuries. Her only joy was tending
the tiny vineyard of fine purple grapes that climbed
the slope outside her small, two-room, stone house.

One autumn afternoon, as Mrs. Udvari napped, a
low rumbling sound and a fearsome shaking jolted

her from sleep. She ran to the window of her trembling house just in time to see her precious vineyard slide down the hill toward the river.

Even before she could grasp what had happened, the floor of her home slipped from under her feet, the walls tumbled around her, and she and the shattered stones slid down the hillside after the grapes.

Several minutes later, bruised and battered, Mrs. Udvari pulled herself from the wreckage of her house and discovered that she had fallen more than 25 yards down the slope. And so had her beloved vineyard. Incredibly, the entire vineyard sat beside her — intact!

All the vine props still stood straight, each row arranged just as neat as before. Indeed, not even a single grape seemed out of place.

Mrs. Udvari and her vineyard had been caught in a very unusual landslide that occurred on September 15, 1971, at Dunafoldvar, just south of Budapest on the western bank of the Danube.

Other minor landslides had occurred near here often, for the river banks are high and steep and their soil is soft and rich.

The 1970 landslide was bigger than any in recent history and had all the earmarks of a major earthquake. More than a million cubic yards of earth slid into the river. A tremendous dust cloud rose over the area. And violent waves washed back and forth across the river.

When the dust settled and the waters calmed, people found that two new islands, both long and narrow, had been created in the river running parallel to the bank.

Naturally, everyone assumed that the islands had

been formed by debris sliding into the water from the high bank above. But later research proved this wrong. Instead, something very odd had happened at Dunafoldvar.

Both islands were approximately 900 feet long and some 30 feet wide. Their surfaces were split by great cracks and studded with crazily tilted blocks of clay bigger than a man. They looked like small mountain ranges poking up from the water.

More puzzling, however, was the discovery that the islands seemed composed of the same material as the riverbed. In fact, shells, snails, and fish carcasses dotted their surfaces. Obviously, these new islands had not slid into the river from above, but had been pushed up from below!

Investigations showed that torrential rains had fallen for several weeks before the slide and probably loosened the earth beneath the surface. This substrata apparently started sliding downhill, at first leaving the top soil undisturbed. Because most of the movement was in the lower depths of the riverbank, the ground beneath became compressed, thus forcing the riverbed to push upward out of the water and create the islands.

For Mrs. Udvari, this strange landslide produced one happy side-effect. Although her stone house couldn't withstand the shock, her well-rooted vineyard held together. Anchored in a layer of top soil, the vineyard simply rode down the riverbank on top of the landslide to its new location without any damage.

CORNELL

BLOW-OUT ON PLATFORM "A"

THE DRILLING CREW on Union Oil's Platform A, five and a half miles off the Pacific Coast, had just bored past the 3,000-foot level when the bit broke into a undersea cave and hit a gigantic natural gas deposit.

Under pressure, the gas blew back through the drilling hole, spewing gas and crude oil over the men and machinery on the platform and sloshing back into the sea. This was the start of the most publicized ecological disaster in American history — the Santa Barbara, California, Oil Spill.

Beginning on January 28, 1970, and continuing for the next ten days, oil and gas gushed from the sea bed at a rate estimated near 10,000 barrels a day. The drilling crew finally capped the leak with a cement plug, but not before some 800-square miles of ocean had been covered with a black, sticky, oil slick.

The actual amount of oil that flowed from the Platform A leak may never be known. Estimates vary widely, depending on whether you talk with an oil company representative or an ecologist. But the well continued to leak at least small amounts for more than 12 months.

Evaluations of the oil pollution's effect on the environment also vary widely. According to the oil men, there had been little damage. Conservationists, on the other hand, claimed death and destruction to wildlife was catastrophic. Even later and more detailed studies by scientific teams of marine biologists seem to contradict each other.

Some facts, however, are undeniable. Scores of dead sea birds floated to shore immediately following the well blow-out, almost certainly the victims of the oil pollution. As one biologist explained: "Any birds that settle in the oil slick surely must die, because they naturally preen themselves to get the oil off their wings. This means they take in some of the oil, which is poisonous."

Newspaper reports claimed as many as 15,000 birds had died. Although this figure was never proved, at least 1,500 birds were known to have been treated by volunteers who attempted to wash them clean of oil. Among the many species killed by the pollution were: cormorants, loons, common grebes, scoters, marbled godwits, willits, dowitchers, western gulls, California gulls, and Heerman gulls.

Marine biologists are not so certain about the effects on other members of this ecological system. The oil killed outright many forms of marine life, such as the abalone, periwinkles, and rock crabs. Yet, other forms seemed to survive despite the heavy oil coating, especially the mussels and sea anemones. In addition, fish eggs and larvae, as well as the plankton population (the tiny green plants), showed no ill effects.

Only a few sea mammals — seals, whales, and dolphins — died from suffocation when oil filled their blowholes and nostrils. Fortunately, most mammals have sense enough to avoid oil slicks. As one biologist noted, "When whales see discolored water, they swim far around it."

Ironically, more sea creatures may have been killed by man's clean-up efforts than by the oil. Corexit, a chemical used to disperse the oil by breaking it up in small droplets that fall to the ocean bottom, is a very poisonous detergent.

The overall, long-term effects of the oil spill may not be known for many years. In the meantime, the controversy over the damage rages on. Of course, there is one point that no one argues about. The leaky well contaminated at least 100 miles of beautiful sandy beaches with an ugly, sticky, smelly dirty scum of black oil.

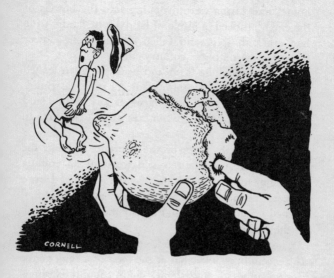

A PUSHMI-PULLYU
EARTH

SOMETIMES strange changes in the earth's surface
happening halfway around the world from each
other seem oddly linked together.

For example, while the natives of Ticrapo, Peru,
watched their little city slowly sink into the earth,
people half a world away on the Philippine Island
of Davao Oriental watched a huge bubble of dirt
bulge out of the ground.

Messengers from Ticrapo brought an incredible tale to government officials in the capital. Their homes and farms were dropping into the bowels of the earth! Indeed, a circle of land about 1 mile wide had cracked, split, and fallen several inches lower than the surrounding countryside.

The simple adobe farm houses of the town leaned at crazy angles like stubby Towers of Pisa. Buildings appeared ready to collapse at any second and the farmers moved their families and possessions to safer ground.

Almost within days of the Ticrapo sinking, the Smithsonian Institute received word of an equally odd happening near Baganda on the island of Davao Oriental in the Philippines. A huge earthmound had pushed up out of the ground near the northern bank of the Mahanob River.

The bulging nob of land, some 150 feet long and 15 feet high, was located at the base of a long gentle ridge leading to the island's main mountain range. Large cracks appeared along the ridge and a small stream that once flowed beneath it had risen out of the ground so its dry bed traced a path up the side of the hill.

The dramatic rising of the earth had happened quickly, popping up some dozen feet in less than three weeks. At the same time, earthquake tremors shook the region and the sound of crushing or sliding rocks could be heard.

What was the connection between the land rise at Baganda and the land sink at Ticrapo thousands of miles away? None at all, said most scientists. Although they agreed the earth is really a plastic body constantly changing shape, this case of a simultaneous push and pull on the planet's surface was only a coincidence.

The land rise in the Philippines was a phenomena known as "slumping," that is, the underground movement of dirt and boulders due to erosion. If the subsurface landslide is suddenly stopped, the dirt creates a weird pile-up that looks like a bulge in the ground.

The Ticrapo sinking, on the other hand, probably was due to a weak earthquake in which the land simply shifted and settled along a major crack in the earth's crust.

Coincidence or not, the rise and fall of the earth on opposite sides of the world almost looked as if some playful giant had poked his finger into the ground at Ticrapo and watched it poke out at Baganda on the other side of the planet.

THE BLOODY TIDE

ALL ALONG the Louisiana coast in June 1969, fishermen, off-shore oil riggers, and beachcombers watched in amazement as the water turned red almost before their eyes.

As the tide rolled into the bayous of Terrebonne, La Fourche, Jefferson, and Plaquemines parishes, the water swirled scarlet around cypress trees and lapped pinkly against beaches.

Then, right behind this strange red tide, came wave after wave of dead fish.

According to estimates by the Louisiana Fish and Wildlife Commission, more than 125 tons of fish, including menhaden, trout, crab, shrimp, and sting rays washed up along a 120-mile stretch of coast west of the Mississippi Delta. All were apparently victims of the mysterious red tide.

Actually, the red tide was no mystery to biologists. "Red tide" is the popular name for a rare biological event that is brief, spectacular, and deadly to underwater inhabitants.

The water really turns red from the flowers of a tiny, almost microscopic, aquatic plant that usually blooms in the late summer or early fall. As it blooms, the plant not only colors the water, it releases a poisonous substance deadly to fish.

Ironically, the organism is itself extremely fragile and deteriorates shortly after it blooms. But not before it has done its deadly job.

Massive killings of fish from other red tides have been reported occasionally along the Gulf of Mexico coast. (A similar red tide off the shores of Peru in December 1970 killed literally millions of fish!)

The Louisiana red tide came early and unexpectedly. Apparently weather conditions, water temperature, and even mineral content of the sea were perfect for promoting the bloom. Other aquatic plants have similar life cycles and may also cause multicolored tides, but only a few produce the toxic substances capable of killing.

Unfortunately, there is no way to predict — or prevent — the more deadly forms of the tide. The men who watch the waters along the Gulf Coast only know that when the water runs blood-red, death usually follows in its wake.

HUNGRY GYPSIES

Moths have been the enemy of housewives and haberdashers for centuries. Who knows how many millions of curtains, blankets, coats, and blue serge suits have been destroyed by moths?

Certain moths, such as the Gypsy Moth, are also a plague to foresters and lumbermen. The hungry gypsies destroy millions of dollars worth of woodland each year. When thousands of moth larva, or caterpillars, attack a forest, they can defoliate it almost as fast as any chemicals used in Viet Nam.

Once a tree loses its leaves, it becomes vulnerable to weather conditions. Indeed, a tree attacked by Gypsy Moths rarely makes it through the next winter.

When trees die, the homes of other species of wildlife are destroyed, soil erosion increases, and the forest fire hazard grows. Without roots to hold the soil, even the course of streams and rivers can be changed.

In short, the Gypsy Moth is a giant pest — and an unnecessary and destructive link in the ecological chain of life.

During the summer of 1969, infestations of Gypsy Moths defoliated more than 150,000 acres of trees in Connecticut, New York, New Jersey, and Pennsylvania.

Alarmed by this sudden destruction, the Pest Control Division of the U.S. Department of Agriculture launched a full-scale attack on the pesty insects.

Of course, the easiest method of controlling the Gypsy Moth is with pesticides. But the growing concern over the side-effects of DDT has caused officials to seek new, natural means for controlling troublesome pests.

For example, the Gypsy Moth has at least seven different natural enemies. One small wasp from India attacks the eggs of the moth. Other insects from Spain eat the larva stages. The Carabid beetle eats both eggs and larva.

A spray has been developed that interferes with the caterpillar's ability to eat. And a special virus that attacks only the Gypsy Moth is being tested.

Perhaps, the best — but most complicated —

means of dealing with the moths is the "sterile release method" — a sort of birth control for bugs. In this form of planned parenthood, massive numbers of moths are raised in special nurseries. The male moths are then sterilized with gamma rays and released into the wild forests. The presence of so many sterilized males naturally reduces the size of the next generation.

The Gypsy Moth's days of destruction may soon be coming to an end. And, for once, man may even have found a way to kill off the pest without killing the rest of us at the same time.

LIFE IN SPACE?

THE FARMHANDS cursed the early morning heat of
the Australian spring as they stacked hay in a barn
near Murchison, Victoria. Suddenly, a low rumbling
noise like thunder shook the barn, although no
storm clouds appeared in the sky. Seconds later, a
small dark stone crashed through the barn's roof
and narrowly missed striking one of the farmers.

The men picked up the cool smooth rock. It was

unusually heavy for its size and covered with a thin, black crust as if scorched in a fire. One edge of the rock had broken off, revealing a grayish cementlike interior flecked with bright metal dots. Most strange, the stone gave off the peculiar odor of denatured alcohol.

Puzzled and perplexed, the men tried to explain how this unusual rock had come hurtling into their hayrack. Little did they know that they held a meteorite — a piece of cosmic debris probably left over from the breakup of some ancient planet. Nor could they imagine that this little stone from the sky would produce one of the most exciting scientific discoveries in history.

More than thirty fragments of what is now called the "Murchison Meteorite Fall" eventually were recovered near this Australian town. Most fell on a dirt road, but several others — like the barn stone — literally fell in the laps of some men standing around the concrete parking lot of a neighboring dairy.

The meteorites fell after a spectacular fireball streaked over southeastern Australia in broad daylight on September 28, 1969. Thousands of people watched this giant meteor with its long trail of smoke and dust enter the atmosphere and explode in midair.

Learning of the fireball and the recovered meteorites, the Smithsonian contacted Australian officials and arranged for samples to be distributed to research laboratories in the United States and elsewhere.

Scientists quickly identified the meteorites as carbonaceous chondrites, a rare type of meteorite rich in carbon materials and long thought to contain some clues to the origin of life in the solar system. Unfortunately, no real evidence had ever been found to prove this theory.

But this time, the scientists found their proof.

Little more than a year after the space stones crashed on earth, scientists at the National Aeronautics and Space Administration announced they had detected the presence of amino acids in the meteorites. Amino acids are complex organic compounds that form the basic building blocks of life, and this was the first time they had ever been found in a meteorite.

That little black rock found in the Australian hayrack may prove to be one of the most important pieces in the intricate puzzle of how life began.

THE BURNING SOIL

FARMERS walking to their gardens in western Samar one morning in 1969 noticed that the ground beneath their feet seemed unusually warm.

The weather in this part of the Philippine Islands had been extremely dry and hot for several weeks. Still, the ground couldn't get that hot from the sun!

Then, as the farmers stared in disbelief, thin wisps of smoke drifted up between their toes.

The ground was on fire!

The Philippines lie on one of the great "rings of fire" that encircle the Pacific. Volcanic activity in this area of the world is very common and horribly violent. The terrified natives of Samar thought the burning soil certainly must be a warning of some terrible eruption about to occur.

Yet, for many days afterward, the fields of Samar continued to smolder without any volcanic eruption. The lush, green covering of grass and reeds turned brown, then black, and then finally died. Gigantic coconut trees, their roots burned through, tumbled to the ground with a crash, sending up little clouds of ashes and burnt twigs.

But still no eruption. The land simply smoked. The ground became covered with a thin layer of ashes. And a veil of white smoke hung over the island farmlands.

Finally a team of scientists from the Commission on Volcanology arrived from Manila for an inspection tour. They found no evidence of volcanic activity, but their discovery was even more unbelievable.

They found the soil at Samar as inflammable as a Molotov cocktail!

The Samar dirt is black, lumpy, porous, and filled with hydrocarbons. In other words, the soil is really a mixture of dirt and vegetable matter much like the ancient swamps. In many parts of the world, this rich earth was pushed deep beneath the surface eons ago to become underground deposits of coal and oil.

At Samar, this mixture lies close to the surface. It is just like old oily rags — ready to burst into flames under the right conditions. The long dry spell in Samar had made the ground highly combustible, so that a campfire or a carelessly discarded match could touch off an underground blaze.

Apparently the "Samar soil burn" had started in some cultivated land and then spread slowly and undetected beneath the surface until it engulfed more than sixty acres of land. Over 35,000 coconut and other fruit trees were destroyed before the underground fire burned itself out.

THE SEA SURGE AT VALPARAISO

SOMETIMES physical changes in the earth occur so far away from human settlements that they go unnoticed and unrecorded. Yet, all events in nature are interrelated and every action has a counter reaction.

On July 25, 1968, unusually high waves began rolling into shore along the coast of Chile. Waves more than 18 feet high swelled into harbors. At

Valparaiso, ships bobbed wildly, broke their moorings, and crashed into docks.

Thousands of little fishing boats tossed and bucked in the churning surf, frightening their superstitious crews who had seen no signs of an approaching storm. The waves struck suddenly and without warning and then ceased as quickly as they had begun.

High waves along the coast of Chile are often caused by northerly winds, but there had been no breezes that day. Nor could scientists record any off-shore earthquake activity that might have triggered these near-tidal waves.

What was the source of these mysterious and murderous waves? Why had the sea surged forth at Valparaiso?

Most experts agree the waves could only have been created by some great underwater disturbance somewhere far out in the Pacific. Perhaps thousands of miles away, on a barren and deserted atoll, a volcano had erupted in terrible fury. Or maybe an earthquake caused by the spreading of the sea bed along one of the great cracks in the ocean floor had churned up the waves that would some later day strike Chile.

Earthquakes in the Philippines and volcanoes in the Fiji Islands had caused similar tidal waves in the past. Unfortunately, this time, no report of unusual activity had been received from any part of the Pacific.

The exact cause of the sea surge at Valparaiso has never been determined. Somewhere, sometime, at some unknown spot in the vast Pacific, the waves had begun rolling eastward toward Chile. Days, weeks, maybe even months later, these waves crashed, unexpected and unpredicted, against the shore.

The waves were man's only evidence that something strange and powerful had torn apart the ocean's floor many, many miles away.

A MODERN BIBLICAL PLAGUE

"AND THEY SHALL COVER the face of the earth, that one cannot be able to see the earth . . . and shall eat every tree that groweth for you out of the field."

This quotation from *Exodus* describes the plague of locusts called down on Egypt by Moses.

Unfortunately, the great plagues of locusts are not just Biblical tales. Even today, locusts are still dreaded threats to farms and fields around the world. Year after year, in the hot, dry lands bordering the Red Sea, these horrible locusts take their toll in valuable, vital food crops.

In 1968, one of the most terrible plagues in recorded history struck the ancient kingdoms of Egypt, the Sudan, Ethiopia, Somaliland, and Eriteria.

Five Sudanese provinces with a total of more than 200,000 square acres were infested with giant swarms of the hopping, flying, chirping insects. This was almost twice the area attacked by any previous plague in modern times.

Poor farmers watched helplessly as the sky turned black with clouds of flying locust. Literally millions of the hungry pests settled down on fields and trees, devouring everything in sight.

At times the locusts were so thick they even seemed to swallow the air from the sky. The Smithsonian received reports of individual swarms more than six miles wide.

The plagues of the Sudan, Somaliland, southern Egypt, and the other African countries were particularly tragic. The land is generally dry and barren, with most crops concentrated in the narrow strip of oasis land along the Nile River. Millions of natives in this region already lived near the starvation level, even without the added curse of the locusts.

For these primitive simple people, life is still little different than it was in Biblical times — including even the famous plagues that scourged the land of pharaohs and pyramids.

FOOTPRINTS FROM
THE PAST

ONE FINE DAY in January 1969, a power-shovel
operator left his home in suburban Sante Fe,
New Mexico, and drove to a construction site at the
edge of the city's airport — just as he might have
on any other day. He carried a lunch packed by his
wife and he listened to the car radio on his way to
work — just like any other day. And — just like any
other day — he didn't expect his work to be any

more exciting than clearing and loading some cinder piles.

Little did he know that this day he would uncover history!

One day, some 100,000 years before the shovel operator left home, a giant North American camel took a stroll across a mesa that would someday become the site of Sante Fe's airport.

Under a slate gray sky dappled with brooding black clouds, the camel walked unperturbed past a score of tiny, hissing steam geysers, around some bubbling pits of boiling mud, and by a cluster of small but very active volcanoes.

With single-minded determination, this huge humpless long-necked ancestor of today's camel headed for a patch of green vegetation on the horizon.

On his way to this dinner, the camel trod straight through a shallow puddle of cool mud that oozed around his oversized hoofs. Behind him stretched a trail of clear deep prints marking his path across what would someday be an airport runway.

Shortly after the camel passed this way, the cluster of volcanoes began one of their periodic eruptions. A mass of cinders, ash, and lava spewed up, baking the footprints solid in the mud and covering them under a deep layer of cinders.

Thousands of years later, while clearing these same cinders and loading them on a truck, the shovel operator from Sante Fe noticed a strange pattern in the hard-packed clay. He stopped work and called his foreman, who called airport officials, who, in turn, called scientists at the Museum of New Mexico.

A team of scientists arrived within the day to

measure, photograph, and make plaster casts of the tracks. The area was roped off and a steady stream of tourists and curiosity seekers soon lined up to see this prehistoric oddity.

Perhaps, as they looked at these footprints from the past, some of these modern men may have wondered if their own tracks in time would ever be unearthed by some future civilization.

ONE MORE VICTIM?

The Hawaiian monk seal is one animal that cannot stand the sight of man.

As man pushes out from the overcrowded mainland into the undeveloped islands of the Pacific, he is apparently driving the man-hating seals to extinction.

One scientist with the U.S. Department of Interior, who has studied the habits of the monk seal for several decades, reports many of the highly developed islands, such as Midway, have lost almost their entire seal populations since the late 1950's.

On islands such as Kure and Green, where man has only recently arrived, the seal population has already dropped by some 50 percent.

The Hawaiian monk seal has always avoided beaches where they might be disturbed by man. Unfortunately, the number of beaches where the seals can still hide from man is dwindling fast.

Man's very presence at a rookery (breeding ground) seems to have a very unhealthy effect on the seal young. Although the reason is not well understood yet, baby seals on inhabited islands have a much lower survival rate than those born on deserted islands. As many as 50 percent of the seal pups die when man is nearby. At this rate of infant mortality, the seal herds cannot reproduce young fast enough to balance off the normal death rate of the older seals.

Worse yet, the Pacific seals don't seem to be getting the message that man is bad for them. At least, there has been so far no mass migration of seals to uninhabited atolls. Instead, they stay on at the old rookeries and slowly die off.

The tragic inability of the Hawaiian monk seal to live with man seems to be a characteristic of all seals. Certainly, it explains why no seals breed among the main islands of Hawaii; or why seals are extremely rare in the Mediterranean; or why they have disappeared completely from the Caribbean.

The world's population of Hawaiian monk seals is now disappearing at an estimated 20 percent a decade. Unless the United States establishes some natural game preserves — small atolls and islands where the seals can be alone — the Hawaiian monk seal could join the passenger pigeon and the dodo as still another victim of man.

MARAUDING MICE

AUSTRALIA is famous for its unusual and unique creatures — the kangaroo, the wombat, the koala bear, the duckbilled platypus. But in recent years, the biggest animal newsmaker "Down Under" has been the *mus musculus*.

The *mus musculus* is better known to housewives from Sydney to San Francisco as the common house mouse. This tiny speed demon with a giant appetite can make a shambles out of any kitchen pantry overnight.

In Australia, the house mouse lives outdoors and feeds on plants and grains. In the past several years, the mouse population of Australia has grown to plague proportions.

In March 1970, Australian farmers in the states of Victoria and New South Wales reported that millions of the little rodents were sweeping through the area eating everything in their paths.

The town of Cuyen, 200 miles northwest of Melbourne, was reported "alive" with mice. In other parts of the country, so many mice covered the highways that speed limits were reduced to 20 miles per hour to prevent skidding on their bodies.

In some hospitals, patients had to be protected from mice attacks by placing inverted tin cans around the legs of their beds. At the town of Sea Lake, police reported the mice formed a "moving carpet" that gobbled up grapes, garden vegetables, and even other dead mice. At Hopetown, farmers trapped as many as 300 to 400 mice in a single night.

The most serious effect of the mouse plague was the damage to grain crops. Rough estimates made in the fields indicated the mice had eaten up more than 20 percent of the year's rice, corn, and wheat.

All attempts to halt the mice seemed futile. Flooding the fields only drove the mice to drier ground. Poisons backfired and started killing livestock and pets as well.

About the only thing that stops the mice is cold weather, when large numbers die off from exposure and starvation.

Mouse plagues are not unknown in Australian history, and particularly bad infestations occurred in 1917 and 1938. The great increase in the mouse population usually follows a number of years of high rainfall. The recent plague came after three years of unusually rainy weather.

Obviously, the man who could build a better mousetrap would have a path beaten to his door in Australia. Or, maybe they just need a Pied Piper "Down Under."

THE FATAL FLOWERS

WHEN THE WOODLANDS of America burst into bloom each spring, this country rejoices in the beauty of nature. When the bamboo forests of Japan begin to flower, that country goes into mourning.

No, this Japanese despair over the blooming bamboo is not just a strange Oriental custom. Bamboo is a very valuable commodity in Japan; and once the unusual ma-dake species of bamboo flowers, it dies.

The ma-dake bamboo is a type of "century plant" that blooms only once every hundred years. (Actually, the cycle runs between 60 and 120 years, depending on the average temperature of the area where it grows.)

This fatal flowering usually affects entire forests at the same time. In other words, all the plants of the same "generation" that are born together, live, flower, and die together too.

The blooming of bamboo in a particular region usually lasts from one to fifteen years. But the end result is always the same. About a year after the flowers appear, the greenish-yellow stalks wither.

The latest cycle of blooming and dying began in 1960, when about 30 percent of the Japanese bamboo crop burst into blossoms and then wilted away. The peak of the current flowering came in 1969, when almost 50 percent of the remaining bamboo forests were affected.

Unlike other plants, the flowers of the bamboo produce no fruits or seeds. Instead, the bamboo reproduces itself by sending out new roots.

The roots are usually unaffected by cutting or breaking healthy bamboo. But the flowering death kills even the roots. Almost ten years is required for new roots to take hold and to produce new and harvestable crops after the blooming.

Obviously, in Japan where bamboo is used for construction, paper-making, art, and thousands of other purposes, the once-a-century appearance of those deadly flowers is catastrophic.

Imagine how Americans would feel if the pine, evergreen, and redwood forests wilted and died every one hundred years.

THE FLOATING ISLAND

THE NAVY LOOKOUT on the destroyer escort *John D. Pearce* couldn't believe his eyes.

Out there, in the middle of the Windward Passage, halfway between Cuba and Haiti and 60 miles from any land, a tiny island was floating along at two and a half knots.

Floating islands have been part of sea lore since the Phoenicians first sailed out into the Atlantic.

Some of the legends have a basis in truth, for large chunks of coastline do often break loose and drift out to sea. The matted undergrowth and roots help hold the soil together and keep it afloat.

The 15-yard-long island spotted by the escort ship was covered with a bushy mangrovelike growth. It also sprouted about a dozen 35-foot palm trees.

Of course, the island had to be reported immediately as a hazard to navigation. But naval authorities were not the only people excited by this odd discovery.

Ecologists at the Smithsonian Institution who heard about the island thought it would make a fine floating laboratory to test some theories about ecology. For example, how do land animals and plants fare at sea? Would new ecological chains develop in this floating environment? Most important, if the island docked at some other body of land, would the plants and animals take up new homes on the mainland?

Perhaps this mysterious island might hold the key to explaining how various species of plants and animals had been scattered around the world.

Seeking answers to these and other questions, a team of scientists rushed from Washington to the U.S. Naval Base at Guantanamo Bay, Cuba, to catch a special island-hunting helicopter.

After catching up with the floating island, the scientists hoped to lower themselves by rope ladder to make on-the-spot investigations and gather samples of the flora and fauna. Later, they'd keep track of the island's drifting, mark its possible landfall, and monitor any cross-cultivation between the island and its new home.

Unfortunately, floating islands prefer to remain

mysteries — a part of the ocean mythology with mermaids and sea monsters.

Even before the scientists boarded their helicopter, the island sank unseen and unmarked somewhere south of Cuba. Any secrets it might have revealed went with it to the bottom of the Caribbean.

THE CORAL KILLERS

A MASSIVE, unexplained, and almost unstoppable
population explosion among a once rare species of
starfish is threatening to destroy the entire ecological
balance of the South Pacific.

For more than a decade, marine biologists have
watched with horror as the ugly, sixteen-legged,

Crown of Thorns starfish has multiplied out of control.

No one would really care about the Crown of Thorns — if only it wasn't so hungry! The starfish's favorite meal is coral, and the creature — which grows up to two feet wide — can devour an area of coral twice its own size in less than twenty-four hours.

Coral is really a tiny animal that secretes a calcium solution. As the calcium oozes out of the coral and hardens into a shell, the coral builds underwater caves that harbor millions of fish and create tough natural breakwaters that protect island beaches.

Once a coral reef is attacked and eaten by the starfish, it begins to crumble and wash away. With the coral reef gone, the fish lose their homes and the beaches become vulnerable to the battering of ocean waves. Eventually, the food supply for millions of people could disappear. Indeed, the very islands of the Pacific could disappear too.

"If the starfish population explosion continues unchecked," warn marine biologists, "the result could be a disaster unparalleled in the history of mankind."

The starfish first appeared in large numbers in 1963 at Green Island, a resort area on the edge of Australia's Great Barrier Reef. By 1969, the Crown of Thorns population had reached incredible proportions. And Australian scientists claimed the starfish were "eating their way from one end of the Reef to the other."

Destruction was not restricted to Australia, however. The starfish invaded the waters around Guam,

Truk, Rota, Johnston, the Fijis, and hundreds of other islands throughout the Pacific and Indian Oceans. At Guam, the Crown of Thorns population jumped from a handful to more than 20,000 in just three years and they ate up nearly 24 miles of coral.

After destroying one reef, the hungry starfish migrate to another. Often the smaller starfish remain behind to feed on any coral that has survived and started to grow again.

By mid-1969, the governments of all the nations touching the Pacific had begun action to halt the deadly starfish. But, although science was ready to fight the plague, no one really knew how it began. Whatever the cause, it probably involved some action by man.

The uncontrolled use of pesticides may have killed off some natural enemy of the starfish. Or, perhaps the widespread gathering of the giant triton shell — the greatest enemy of the starfish — by natives for sale to collectors and tourists may have reduced the number of shells to the point where the starfish could multiply unchecked. Some scientists even think that underwater blasting or atomic tests could have touched off the sudden growth of the Crown of Thorns. Ironically, coral itself is a natural enemy of the starfish, feeding on its larva stage. If enough coral was destroyed by underwater explosions, the baby starfish might have started growing faster than the coral could eat them.

The job of halting the starfish explosion may prove as difficult as finding its cause. The surest method sends down individual divers to kill the starfish with injections of poison. Another method might be to release some new natural enemy of the starfish in the reefs. An important part of any attack will be an education program for the natives to halt the collecting of the giant triton shells, still the best protection against the Crown of Thorns.

The battle to save the coral reefs will be long and costly. Yet, if it isn't fought — and won! — the Crown of Thorns could turn the Pacific into a vast dead sea.

THE MONSTER THAT WASN'T

THE STORY out of Mexico sounded like a rerun of the *Creature from the Black Lagoon*.

According to United Press International, a 35-ton, 30-foot, unidentified sea creature had washed up on the beach at Tecolutla.

The creature's serpentlike body was covered with armor plate and jointed so it could swim. Most amazing, the so-called monster had a ten-foot horn protruding from its head.

The breathless report speculated that the creature might be a survivor from the age of dinosaurs somehow preserved in an iceberg for millions of years. Or, the story suggested, the creature might have been driven from a dark, deep submarine cave by recent underwater blasting in the area.

This scanty report was both unbelievable and intriguing. In the early 1950's, African fishermen netted a specimen of fish thought extinct since prehistoric times. Could the Mexican sea monster be another of these lost animals?

Skeptical, but hopeful, officials at the Smithsonian Institution sent off urgent telegrams asking for more information.

Within hours, a telegram came back from a Mexican biologist who claimed to have investigated the carcass personally. He confirmed the general size and shape of the creature and reported:

"The animal has some kind of horn approximately 30 inches long and about three inches in diameter. It is not straight, but has a very small curvature — definitely the shape of a horn, not a bone.

"I feel the creature is a kind of mammal and certainly not a fish."

The so-called biologist may have been convinced the creature was not a fish, but the Smithsonian still thought the story sounded a bit fishy. Especially since no one had ever heard of the "biologist" before.

While marine biologists around the world waited for news, the Smithsonian asked the Biological Science Institute at Tampico, Mexico, to check out the monster.

Finally, several days later, the Smithsonian received a cable from a team of well-known scientists at the Institute.

The monster was nothing more than a common whale!

Apparently, the ill-fated creature had been attacked by sharks and its partly eaten body washed up on the shore. The horn was not a horn at all, but only a part of the skull that had fractured and split from the main bone.

The sea monster story had been all an elaborate hoax. Superstitious natives — and an imaginative reporter — had turned the grotesque decomposing body into a mysterious monster.

The so-called "biologist," who first confirmed that the creature and its horn were real, was never heard from again.

For a few days, however, marine biologists had held their breath — hoping for something truly unexpected and unbelievable. Perhaps some day a real creature from the past will be discovered. It is unlikely, of course, but science-fiction fans and scientists, too, still keeping hoping.

LIVING CAVEMEN

STONE AGE PEOPLE still living in the twentieth century? Sounds impossible, yet as recently as 1969, a lost tribe of primitive Indians, thought to be extinct, was rediscovered in the jungles of South America.

The Akuriyo Indians of Surinam may be some of the most primitive people left in the world. They still live the way cavemen did thousands of years ago.

A survey party charting the rugged forest area where the borders of Brazil, Surinam, and French Guiana meet, first discovered these forgotten people in 1937. More than thirty years passed before Ivan Schoen, a Protestant missionary, finally tracked them down again.

Schoen described the Akuriyos as a short and stocky people with a pallid, sallow complexion, apparently from living so long in the darkness of the deep forests. He also described them as incredibly filthy, primarily because they camp in the middle of muddy swamps — the only flat and open spaces in this thick jungle country.

The Akuriyos used only the crudest tools and weapons, most of them no more than sharpened stones. The men wore nothing more than skimpy loin cloth of woven fibers; the women only a short apron of seeds and fruit pits on strings.

Naturally, on this first contact, the Akuriyos were suspicious of Schoen and his white companions. He decided not to try and communicate with the Indians. Instead he returned to the coast, leaving behind five Trio Indian guides who he hoped could make friends with the Akuriyos and learn their ways.

Unfortunately, the Akuriyos proved crafty cavemen. They stole the possessions of the Trios, ate all their food, and managed to lose them on the tricky jungle paths.

Schoen's return visit with the Akuriyos was more successful. Although the tribe still treated the white men with suspicion, they invited another group of Trio Indians to stay on as friends. This time, Schoen left a shortwave radio with the Trios so they could rely back information about the tribe's life-style.

The Trios' information confirmed that the Akuriyos were the same Indians found in 1937. In fact, one old man still carried part of a machete given him by the survey party.

Schoen also found that the Akuriyos were a rootless, wandering people who lived entirely by hunting, fishing, and picking ripe fruits from the forest. They had no agriculture and carried no possessions except for their weapons and some simple toys for the children. Because of some ancient taboo, they avoided contact with both white men and other Indians.

The Akuriyos seemed almost compulsive wanderers, roaming over some 10,000 square miles of jungle. Since this was neither very rich nor fertile land, they needed to travel far and wide for game. Yet, it was not the need for meat that drove them. It was the taste of honey!

The Akuriyos were *honey addicts*, with an insatiable desire for the sticky sweet. Even the crude, short-handled stone axes they carried were called "honey-getting things." The best axes were those that could most easily rip open a honey tree.

Of course, these stone axes also marked the tribe as primitive and prehistoric. Most other tribes in South America have progressed far enough to use the metal tools bartered from traders at the edges of civilization. Thirty years after their first contact with the white man, the Akuriyos still clung to their crude stone axes.

Ivan Schoen told the Smithsonian that only 60

or 70 Akuriyos still exist today. How much longer they will exist is a question. Civilization is fast closing in on these modern cavemen. The Surinam government is attempting to open this last wilderness area for settlement and agricultural development.

The Akuriyos — perhaps the last human relics of the Stone Age — may finally become part of the twentieth century. Or, perhaps, they will simply disappear again — this time forever.

INVASION OF THE
GIANT SNAILS

The people of Miami, Florida, thought they had become victims of some horrible science-fiction nightmare when their pleasant, sunny residential neighborhoods were invaded by giant snails — some as big as a man's hand.

Perhaps "invasion" is not the right word. The snails actually were *brought* to Miami by some unsuspecting tourist who spotted the unusual shelled creatures in the tropics and thought they might make nice souvenirs.

Unfortunately, the giant snails — known as African snails or *Achatina fulica* — have plenty of natural enemies in other parts of the world, but few in Florida. Unchecked by normal forces, the snails began reproducing at a fantastic rate.

Because their appetites match their size, the giant snails created a giant problem for the Florida Department of Agriculture. The officials had visions of the snails gobbling up every bit of green vegetation in this famous resort.

Soon after the discovery of this potential snail plague, the state officials launched a massive extermination campaign. Just one week after poison bait was dropped in a 40-acre, area, 2,500 dead snails were picked up. Two weeks later another 5,000 were found. By the end of the year, some 17,000 giant snails had been killed.

The Department of Agriculture didn't stop its extermination campaign even after this killing. Actually, they couldn't afford to stop!

The giant snails often go into a form of hibernation, remaining hidden and alive in some cool spot for more than 6 months. At the same time, the lifespan of these big snails is about five years. And the typical snail mother drops some 400 to 600 eggs a year. Under perfect conditions, then, even one surviving snail could be responsible for producing *11 billion off-spring* in five years.

Why, those snails could have eaten up all of Florida, Georgia, Alabama, and South Carolina and still be hungry!

Luckily, the snails now seem to be stopped. By late 1970, Florida officials reported finding only 5 to 10 snails a week. But the snail watch goes on, just in case the population starts to explode again.

THE VOLCANO THAT FELL IN

TIME seems to have stopped two million years ago on the Galapagos Islands.

This weird world, 580 miles off the coast of Ecuador in the Pacific Ocean, is filled with plants and animals that exist nowhere else, including some throwbacks to prehistoric times.

More than 100 years ago, these bleak volcanic

islands served as the living lab for testing Darwin's theory of evolution. Even today, modern biologists are thrilled by the sight of otter-sized iguanas (land lizards), four-eyed fish, scalesia trees (thirty-foot-high members of the dandelion family), and a host of other rare birds, plants, and insects.

In this topsy-turvey land of ancient relics, everything seems upside down. No wonder then, when the biggest volcano on the islands erupted, it fell *into the earth* instead of out of it.

On June 11, 1968, seismic monitoring stations around the world reported that a volcano on the uninhabited island of Fernandina had erupted with all the force of a multimegaton hydrogen bomb.

A huge mushroom cloud rose high above Fernandina. Many smaller explosions followed the first blast. During the night, volcanic dust in the atmosphere over the island sparked tremendous electrical storms, creating a scene straight out of Hell.

From all appearances, the island looked as if it had been destroyed completely. Scientists everywhere feared that lava flows might wipe out the precious living samples of rare plants and animals.

They were surprised and pleased, however, to find that the volcano had not blown its top like most others. Instead of spewing rocks and lava over the nearby countryside, the caldera (or crater floor) had simply collapsed, dropping almost 300 yards deeper into the bowels of the earth. The collapse caused incredible noise and dust — but little destruction outside the crater.

The first research team reached the rim of the Fernandina Caldera about a week later. The ground still shook with almost continuous tremors.

"The frequency and violence of these tremors was so great," reported one scientist, "that we counted 56 in less than six hours. Each lasted from two to six seconds and caused rock slides all around us.

"Trees shuddered as though a strong wind blew their branches. The tremors rose and fell in waves, so the whole island seemed balanced on top of a jellylike mass."

This continual shaking caused large sections of the crater rim — some more than 1,000 yards long — to fall into the depths. Avalanches of stones fell constantly to the floor of the crater that lay hidden under dust many yards below. The sound of falling rocks was like the roar of heavy seas. And strong gusts of wind rushed out of the crater after each avalanche.

A lake, once located in the northwestern part of the crater, had shifted almost a half-mile to the southeast. A large population of the rare Galapagos ducks had once lived in this little lake. No sign of them could now be found.

Aside from the ducks, however, there seemed few other casualties. Although every part of the island was covered with a light film of gray-black volcanic ash, most of the damage was inside the crater.

Luckily for science, the Fernandina volcano had been an *eruption in reverse*. The crater floor had collapsed from a lack of underground pressure, rather than exploding up and out from a build-up of gas and lava.

In the weird world of the Galapagos — where

118

everything is backwards — this reverse eruption had helped save untold numbers of rare plants and animals that could never be replaced.

Perhaps nature sometimes tries to save its own creatures.

THE CATASTROPHE OF THE CENTURY

No DISASTER this century — not even the great China flood of 1911 that killed 100,000 people or the horrible Tokyo earthquake of 1923 that killed 140,000 — can compare with the cyclone and tidal wave that slammed into East Pakistan on November 13, 1970.

This storm ripped up the Bay of Bengal and hit the unprotected islands at the delta of the Ganges

120

and Brahmaputra Rivers with such sudden and un-believable power that it destroyed 1,000,000 acres of crops, wiped out 235,000 houses and damaged another million, killed 265,000 head of cattle, and took over half a million human lives.

Life is hard on the hundreds of tiny, flat islands in the Bay of Bengal. The soil is rich and fertile, and perfect for growing rice, but the low-lying lands are vulnerable to the slightest change in the weather. Almost every storm blowing up from the Indian Ocean takes a heavy toll of crops — and lives.

The worst storms, known as cyclones, sweep up the funnel-shaped bay and become ever more vio-lent as they are squeezed into the narrow head near the river delta.

The water around the islands is so shallow that waves pushed from the ocean below cannot flow back in deep undercurrents. Instead, the waves wash over the islands.

On November 13, the most powerful cyclone in memory roared up the Bay. Whipped by winds of more than 150 miles an hour, a tidal wave between 15 and 25 feet high smashed into the overpopulated islands.

Striking with full fury in the dead of night, the storm caught most people unaware. The Pakistani government had issued no storm warnings. But even if they had, it probably would have made little dif-ference. Most of the islands have no electricity, no newspapers, no radios, and no television. Indeed, one island had only a single portable radio for a quarter of a million people.

Worse yet, when the storm struck and the sea rose

there was no place to run. None of the islands have hills — or even tall buildings. About the only thing above sea level are the few palm trees.

The storm raged throughout the night, smashing flimsy houses, washing away valuable rice, drowning men and beasts, and sweeping thousands more out to sea.

One man clung to a tree all night as the wind and waters swirled around him. When dawn came, he climbed down to find his farm, livestock, crops, and entire family — including some 25 small children — had disappeared.

Pilots flying over the area reported half of Bhola Island completely destroyed. All the rice on neighboring Hatia Island was washed away. Whole villages had disappeared, with only muddy foundations of buildings marking where they once existed. A huge cargo ship weighing more than 500 tons had been blown 50 yards inland on one island. Not one person remained alive on a string of 13 of the small islands near the port of Chittagong.

On the ground, the disaster seemed even more a nightmare. Bodies hung from trees. Houses lay flattened or gaping open with roofs and walls ripped away. Thousands of dazed men and women — some driven insane by the experience — pawed through the wreckage desperately seeking lost loved ones. Debris from the sea seemed to cover every inch of ground. And the terrible sickening smell of death and decay was everywhere.

The suffering did not end with the dawn, however. Those who survived found food and water supplies either gone or contaminated. And no one came to their relief.

Because the central government of Pakistan is 2,000 miles away on the other side of India, the East Pakistanis traditionally have been treated as poor relations. For example, there was no early warning system for cyclones, even though similar storms 10 years before had killed 65,000 people.

Even now, in this terrible crisis, the government ignored its people. Relief efforts were delayed, confused, inadequate, or nonexistent. When help finally reached the poor people of the islands, it was too little and much too late.

As thousands of islanders died from starvation and disease, resentment against the central government grew in East Pakistan. This smoldering rage burst into revolution in the spring of 1971, when East Pakistan declared its independence.

Unfortunately, the rebellion was quickly and cruelly squashed, and the government seems unlikely to change its attitude toward East Pakistan or to take any steps to prevent a repeat of that terrible disaster.

An early warning radar network to spot the storms and a mass communications system to alert the islands is desperately needed. More important, the government must create "earth platforms," giant artificial hills, on each of the islands so that the people will have some high, dry, and safe retreat should the Bay of Bengal ever rise again.

AFTERWORD:

A HOTLINE FOR SCIENCE

Since 1968 the Smithsonian Institution's Center for Short-lived Phenomena has served as a unique news service providing fast, accurate information about earthquakes, volcanoes, fireballs, animal migrations, oil spills, sea surges, insect infestations, landslides, bird kills, lost tribe discoveries, and a host of other unusual, unpredictable, and often odd-ball events.

More than 2,000 scientists in 50 countries serve as correspondents in the Center's international network reporting news of far-out events in their far-flung corners of the world.

The stories in this book are not fiction. They are real-life events drawn from the files of the Smithsonian Center's headquarters in Cambridge, Massachusetts.

The events retold here occurred without warning and usually caught everyone, except the Center for Short-lived Phenomena, completely by surprise. They were certainly sudden and unexpected, but you'd better believe them.